Union Pacific Railway Company

American shooting association rules for inanimate target shooting 1896

Live bird shooting

Union Pacific Railway Company

American shooting association rules for inanimate target shooting 1896
Live bird shooting

ISBN/EAN: 9783337159726

Hergestellt in Europa, USA, Kanada, Australien, Japan

Cover: Foto ©berggeist007 / pixelio.de

Weitere Bücher finden Sie auf **www.hansebooks.com**

American
Shooting Association Rules

—FOR—

Inanimate Target Shooting, Live Bird Shooting, and
London Gun Club Rules

—AND—

Revised Game Laws

for Western States and Territories.

JUNE 1st, 1896.

CAUTION!

The Game Laws printed herein are corrected up to date. Owing to frequent changes being made at the session of each legislature, we would advise sportsmen and hunters to make inquiry from the State or Territorial officials to learn if any changes have been made.

E. DICKINSON,
Gen'l Manager.

E. L. LOMAX,
Gen'l Pass, & Ticket Agent

OMAHA, NEB.

Unequaled Attractions

ON THE LINE OF THE

UNION PACIFIC

...FOR TOURISTS...

It traverses the Grandest Scenery of the Rocky Mountains, and reaches all the Health and Pleasure Resorts of the Mid-Continent.

SPORTSMEN
Will find in Scores of Localities along this line, Game worthy of their skill, such as....

BEAR, MOUNTAIN LION, COYOTES, ELK, DEER, ANTELOPE, MOUNTAIN SHEEP, FEATHERED GAME OF ALL KINDS.

And Everywhere Are

BEAUTIFUL STREAMS

Well Stocked with Trout.

ANY INFORMATION

Relative to localities for Hunting, or information in regard to the UNION PACIFIC SYSTEM, cheerfully furnished on application to any representative of this Company.

TRAP SHOOTING RULES

OF THE

American Shooting Association.

REVISED JANUARY 1, 1893.

RULES FOR INANIMATE TARGET SHOOTING.

RULE 1. *Judges and Referee.*—Two judges and a referee, or a referee alone, shall be selected by the management, or the contestants, whose decision shall be final.

RULE 2. *Duties of the Referee.*—The referee shall see that the traps are properly set at the beginning of the match and kept in order to the finish. He shall endeavor to make the targets conform to the flight and direction indicated in Rule No. 7. He shall test any trap upon application of the shooter at any time by throwing a trial target therefrom. He may at any time, and must when so requested by a contestant, select one or more cartridges from those of a shooter at the score, and publicly test the same for proper loading. If the cartridge, or cartridges, are found to be improperly loaded, the shooter shall suffer the penalty as provided for in Rule 11.

RULE 3. *Scorer*—A scorer shall be appointed by the management, whose score shall be the official one. All scoring shall be done with ink, or indelible pencil. The scoring of a lost target shall be indicated by a "0," and a broken target by the figure "1."

RULE 4. *Puller.*—A puller, or pullers, shall be appointed by the management, whose duty it shall be to see that the trap or traps shall be instantly sprung when the shooter calls "Pull," and shall be placed in such a position that the shooter will have no means of knowing by his actions which trap is to be pulled. In single target shooting he shall pull the traps as decided by a trap-pulling indicator, or other means that may have been provided by the management, so that the shooter will have no means of knowing which trap the target is to be thrown from.

RULE 5. *Pulling the Traps.*—SECTION 1. Traps may be pulled in regular order from 1 to 3, or 1 to 5, or *vice versa*, if so decided by the management.

SEC. 2. If the shooting is from traps to be pulled in regular order, the shooter may refuse the target from the trap not so pulled; but if he shoots, the result must be scored.

SEC. 3. If the trap is sprung before, or at any noticeable interval after the shooter calls "Pull," he can accept or refuse the target; but if he shoots, the result must be scored.

SEC. 4. If the puller, or pullers, do not pull in accordance to the indicator, or other means provided, they shall be removed and others substituted.

RULE 6. *Arrangement of Traps.*—All matches shall be shot from three to five traps, set level, three or five yards apart, in the segment of a circle (see Diagrams A and B), or in a straight line (see Diagram C). When in the segment of a circle, the radius of the circle shall be eighteen yards. In all cases the shooter's position shall not be less from each trap than the rises provided for in Rule 7. The traps shall be numbered from 1, on the left, to No. 3 or No. 5, on the right, consecutively, according to the number used, as shown in the diagram.

RULE 7. *Adjusting Traps.*—SECTION 1. All traps must be adjusted to throw the targets a distance not less than 40 yards, nor more than 60 yards. If any trap be found too weak to throw the required distance, a new trap or spring that will, must be substituted.

SEC. 2. The lever or projecting arm of the trap shall be so adjusted that the elevation of the target in its flight at a distance of 10 yards from the trap shall not be more than 12 feet, nor less than 6 feet, and the angles of flight shall be as follows:

If three traps are used (see Diagram A): No. 1 trap shall be set to throw a left quartering target. No. 2 trap shall be set to throw a straight-away target. No. 3 trap shall be set to throw a right quartering target.

If five traps are used (see Diagrams B and C): No. 1 trap shall be set to throw a right quartering target. No. 2 trap shall be set to throw a left quartering target. No. 3 trap shall be set to throw a straightaway target. No. 4 trap shall be set to throw a right quartering target. No. 5 trap shall be set to throw a left quartering target.

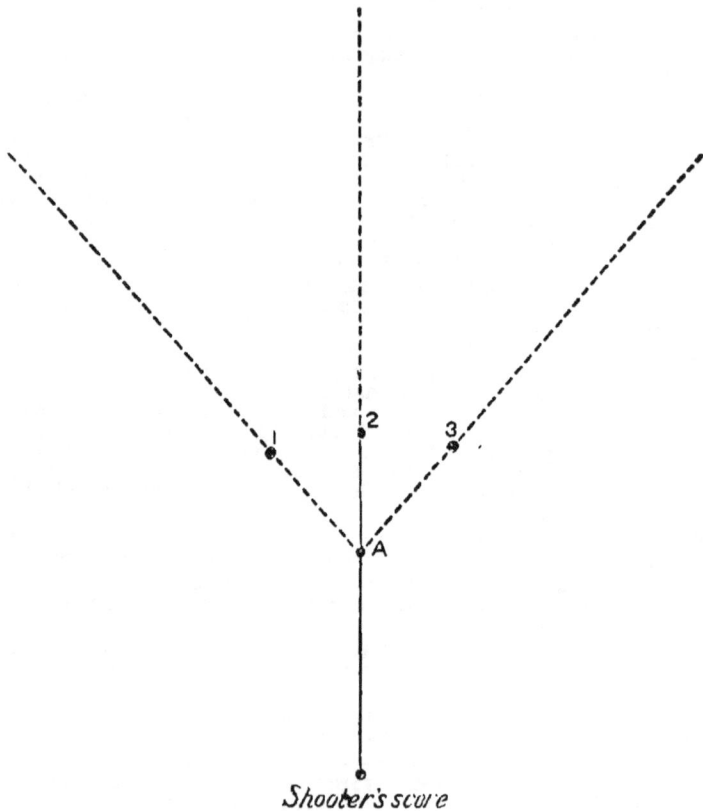

Shooter's score

Diagram A.   (See Rules 6 and 7.)

NOTE.—To get angles for birds thrown from traps 1 and 3, measure six yards from trap No. 2 on line to shooter's score to point marked "A;" lines drawn from this point across traps 1 and 3 will give proper direction of flight.

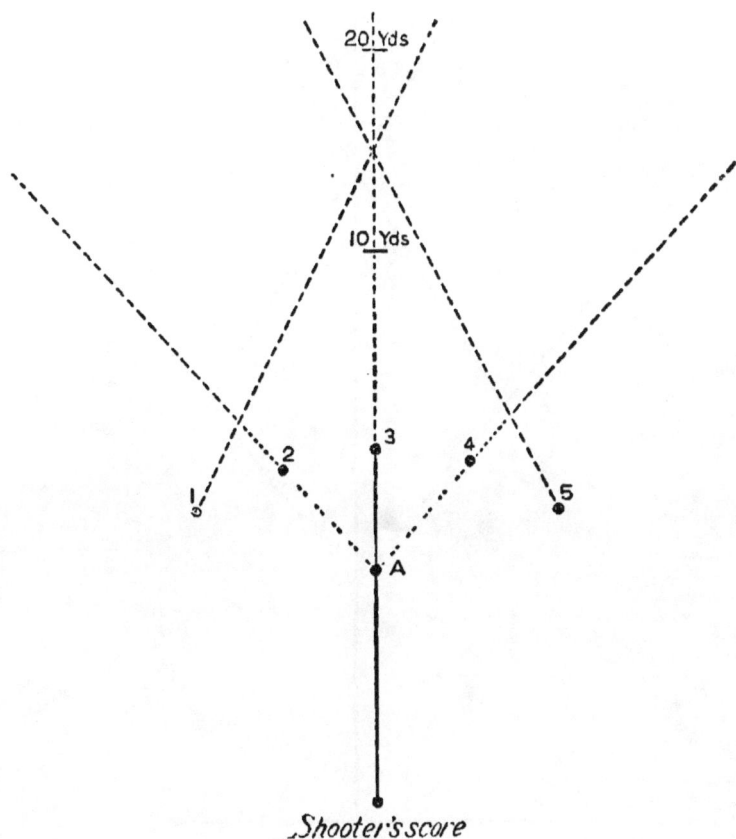

20 Yds

10 Yds

2

3

4

5

A

Shooter's score

Diagram B.   (See Rules 6 and 7.)

NOTE.—To get angles for birds thrown from traps 2 and 4, measure six yards from trap No. 3 on line to shooter's score to point marked "A," lines drawn from this point across traps 2 and 4 will give the proper direction of flight. The birds from traps 1 and 5 should cross the line of flight of the straightaway bird not more than twenty nor less than ten yards from trap 3.

20 ¦ Yds

10 ¦Yds

1   2   3   4   5

'A

Shooter's score

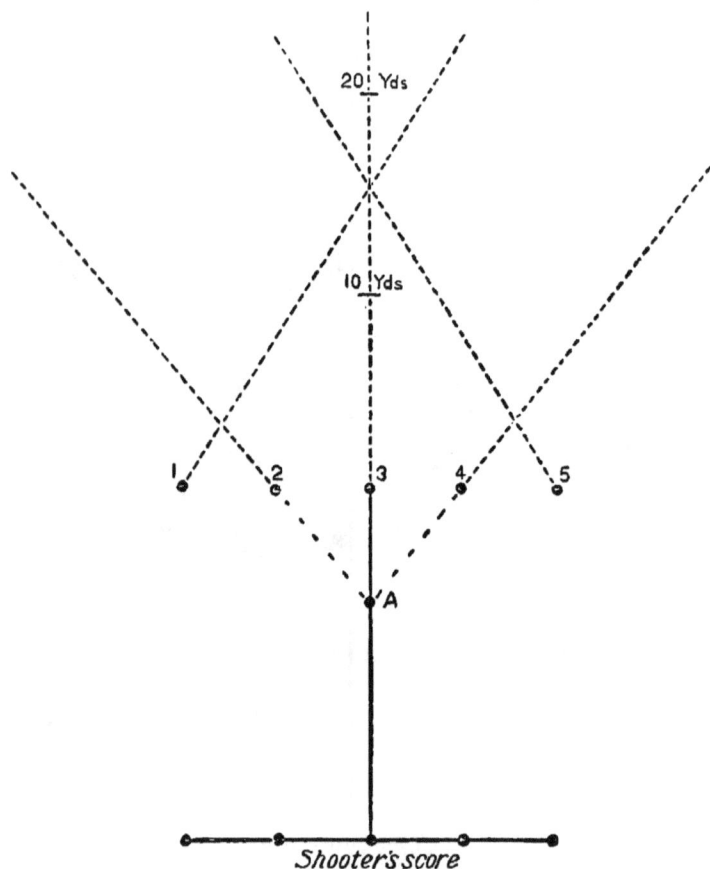

Diagram C.   (See Rules 6 and 7.)

NOTE.—To get angles for birds thrown from traps 2 and 4, measure six yards from trap No. 3 on line to shooter's score to point marked "A;" lines drawn from this point across traps 2 and 4 will give the proper direction of flight. The birds from traps 1 and 5 should cross the line of flight of the straightaway bird not more than twenty nor less than ten yards from trap No. 3.

Traps Nos. 1 and 5 shall be set to throw the targets so that the line of flight shall cross that of the straight-away target at a point not less than 10 yards nor more than 20 yards from trap No. 3.

SEC. 3. After the traps are set for these angles, if the target for any reason shall take a different course it shall be considered a fair target, provided the trap has not been changed.

RULE 8. *Screens* —Either pits or screens, or both, may be used, but the screens must not be higher than is actually necessary to fully protect the trapper.

RULE 9. *The Rise.*—In single target shooting the rise shall be:

18 yards for 10-bore guns.
16 yards for 12-bore guns.
14 yards for 14 and 16-bore guns.
13 yards for 20-bore guns.

In double target shooting the rise shall be:

16 yards for 10-bore guns.
14 yards for 12-bore guns.
12 yards for 14 and 16-bore guns.
11 yards for 20-bore guns.

All distances mentioned in these rules must be accurate measurement.

RULE 10. *Caliber and Weight of Guns.*—No gun of larger caliber than 10-bore shall be used, and the weight of all guns shall be unlimited.

RULE 11. *Loads*—Charge of powder unlimited. Charge of shot not to exceed one and one-quarter ounces American Association, or Dixon's measure, struck. Any shooter using a larger quantity of shot shall forfeit his entrance money and rights in the match.

NOTE—If in the opinion of the management, with the unanimous consent of the contestants, a shooter has not wilfully violated this rule, his entrance money shall be returned to him.

RULE 12. *Loading Guns.*—In single target shooting, only one barrel shall be loaded at a time, and the cartridge shall not be placed in the barrel until after the shooter has taken his position at the score. In double target shooting, both barrels shall be loaded at the score. Cartridges must be removed from the gun before leaving the score.

RULE 13. *Position of Gun* —Any the shooter may adopt.

RULE 14. *Single Target Shooting.*—When the traps are set in the segment of a circle, each contestant shall shoot at three or more targets before leaving the score. If two targets are sprung at the same time, and contestant does not shoot, it shall be declared "No targets;" but if he shoots, the result must be scored.

RULE 15.—*Double Target Shooting.*—Both traps must be pulled simultaneously, and each contestant shall shoot at three or five pairs, consecutively, thrown as follows: If three traps are used, the first pair shall be thrown from 1 and 2, the second pair from 2 and 3, the third pair from 1 and 3, the fourth pair from 1 and 2, and the fifth pair from 2 and 3.

If five traps are used, the first pair shall be thrown from 2 and 3, the second pair from 3 and 4, the third pair from 2 and 4, the fourth pair from 2 and 3, and the fifth pair from 3 and 4.

If only one target is thrown, it shall be declared "No targets."

If a target be lost for reasons stated in Rule 19, Section 1, it shall be declared "No targets." If one be a fair and the other an imperfect target, it shall be declared "No targets." But if the shooter accepts an imperfect target, or targets, the result must be scored.

If both targets are broken by one barrel, it shall be declared "No targets." If the shooter fires both barrels at one target intentionally, it shall be scored "Lost targets." But if the second barrel be discharged accidentally, it shall be "No targets."

RULE 16.—*Rapid Firing System.*—When the traps are set in a straight line and the rapid firing system is to be used, there shall be a screen before each trap on which shall appear the number of the trap, from No. 1 on the left, and each shooter shall stand at score opposite the trap from which the target is to be thrown for him to shoot at. After he has shot at the first target he shall pass to the next score to the right, and so continue until he reaches the end of the score, when he shall return to the score opposite No. 1, and continue as before until his score is finished. If shooters are annoyed, or there is delay in shooting by the smoke of previous shots, the traps may be pulled in reverse order, commencing with the trap on the right.

RULE 17. *Class Shooting.*—All shooting shall be class shooting, unless otherwise stated by the management.

RULE 18. *Broken Targets.*—A target to be scored "broken," must have a perceptible piece broken from it while in the air. A "dusted" target is not a broken target. No target shall be retrieved for shot marks.

If a target be broken by a trap, the shooter may claim another target, as provided for in Rule 19; but if he shoots, the result must be scored.

RULE 19. *Allowing Another Target.*—SECTION 1. The shooter shall be allowed another target for the following reasons:

A—For a target broken by the trap.

B—For any defect in the gun, or load, causing a miss-fire.

C—If the contestant is interfered with, or balked, or there is other similar reason why it should be done, the referee may allow another target.

SEC. 2. When the shooting is at known angles he shall have another target from the same trap; but if the shooting is at unknown angles he shall have another target from an unknown trap, to be decided by the indicator, except it be the last trap, when the shooter has the right to know which trap is to be sprung. In this case he shall have another target from the same trap.

NOTE.—When a shooter in breaking his gun to put in the shells fails to break it far enough to cock the gun, it is considered his own carelessness, and not sufficient excuse for the allowance of another target.

RULE 20. *Lost Targets.*—Targets shall be scored lost if the shooter fails to load, cock, adjust safety on gun, or pulls the wrong trigger.

RULE 21. *Tie Shooting.*—SECTION 1. All ties shall be shot off at the original distance, and as soon after the match as practicable, at the following number of birds:

Ties on Single Targets.—In single target matches of 25 targets, or less, on three traps, 3 targets; five traps, 5 targets. In matches of 26 to 50 inclusive, on three traps, 6 targets; five traps, 10 targets. In matches of over 50, on three traps, 15 targets; five traps, 25 targets.

Ties on Double Targets.—In double target matches of 10 pairs, or less, on three traps, 3 pairs. In matches of more than 10 pairs, 5 pairs, thrown from traps 1 and 3. If five traps are used, the same number shall be thrown in each case, from traps 2 and 4 (unless otherwise arranged by the management, and so stated or understood previous to the beginning of the match).

SEC. 2. If in a series of matches the result prove a tie, such tie shall be shot off at the original number of targets.

RULE 22. *Announcing the Score.*—SECTION 1. When two judges and a referee are serving, one of the judges shall announce the result of each shot distinctly, and it shall be called back by the scorer.

(The call for a broken target shall be "Broke," and the call for a missed target shall be "Lost.")

If the second judge disagrees with the decision of the judge calling, he shall announce it at once before another target is thrown, and the referee shall decide it. In case of another target being thrown before the referee's decision, the target so thrown shall be "No target."

SEC. 2. At the close of each shooter's score the result must be announced. If claimed to be wrong, the error, if any, must be corrected at once.

RULE 23. *Shooter at the Score.*—In all contests the shooter must be at the score within three minutes after his name is called to shoot, or he forfeits his rights in the match.

RULE 24. *Forbidden Shooting.*—No shooting will be permitted in the enclosure other than at the score; and in case there is no enclosure, no shooting within 200 yards of the score, without the consent of the management.

RULES FOR LIVE BIRD SHOOTING.

RULE 1. *Referee.*—A referee shall be appointed by the contestants, or management, whose decision shall be final.

RULE 2. *Duties of Referee.*—The referee shall see that the traps are properly set at the beginning of the match, and kept in order to the finish, and that they are kept properly filled. He may at any time, and must when so requested by a contestant, select one or more cartridges from those of a shooter at the score, and publicly test same for proper loading. If the cartridge, or cartridges, are found to be improperly loaded, the shooter shall suffer the penalty as provided in Rule 15.

RULE 3. *Scorer.*—A scorer shall be appointed by the contestants, or management, whose score shall be the official one. All scoring shall be done with ink or indelible pencil. The scoring of a lost bird shall be indicated by "0," and of a dead bird by the figure "1."

RULE 4. *Puller.*—A puller shall be appointed by the contestants, or management, and shall be placed at least 6 feet behind the shooter, and it shall be his duty to pull the traps evenly and fairly for each contestant, and instantly after the shooter calls "Pull." He must use a trap-pulling indicator, or other device that may be furnished by the management, so that the shooter will not know which trap is to be pulled. All traps must be filled before the shooter calls "pull."

If more than one bird is liberated, the shooter may call "No bird;" but if he shoots, the result may be scored. Should the puller not pull in accordance with the indicator, he should be removed and another puller substituted.

RULE 5. *Arrangement of Traps.*—All matches shall be shot from five ground traps, placed 5 yards apart, in the segment of a circle. The radius of the circle shall be 30 yards from the shooter's score. The traps shall be numbered from No. 1 on the left to No. 5 on the right, consecutively (see Diagram D.).

NOTE.—A ground trap is one that lies flat with the surface of the ground when open, and gives the bird its natural flight in starting.

RULE 6. *The Rise.*—The rise shall be:
30 yards for 10 bore guns.
28 yards for 12-bore guns.
26 yards for 14 and 16-bore guns.
25 yards for 20 bore guns.

RULE 7. *Boundary.*—The boundary for both single and double bird shooting shall be the segment of a 50 yards circle, and a dead line. The circle shall be drawn from a point 10 yards beyond the center trap on a line from the shooter's score, and it shall terminate where it joins the dead line, which shall be drawn at a distance of 30 yards from the center trap, and at right angles with a line drawn from the shooter's score to the center trap (see Diagram D).

RULE 8. *Birds Refusing to Fly.*—When a bird refuses to fly, such artificial means as have been provided by the management may be used to start it, by direction of the referee. A bird hit by a missile shall be declared "No Bird." The shooter may declare a bird refusing to fly when the trap is pulled, "No Bird."

RULE 9. *Gathering Birds.*—A bird to be scored dead must be gathered within bounds before another bird is shot at, and within three minutes' time, by a dog or shooter, or person appointed by the shooter for that purpose. No extraneous means shall de used, and no other person shall be allowed to assist in gathering. If the gatherer cannot locate the bird, he may appeal to the referee to locate it for him. All birds challenged must show flesh-shot marks, to be scored "Dead Birds."

RULE 10. *Birds Killed on the Ground*—A bird killed on the ground with

the first barrel is "No Bird."   But it may be killed on the ground with
the second barrel if the first is fired while the bird is on the wing.   If a bird
is shot at on the ground with the first barrel, and the shooter uses the second
barrel, but fails to kill, it is "Lost Bird."   But if the bird is killed, it shall be
"No Bird."

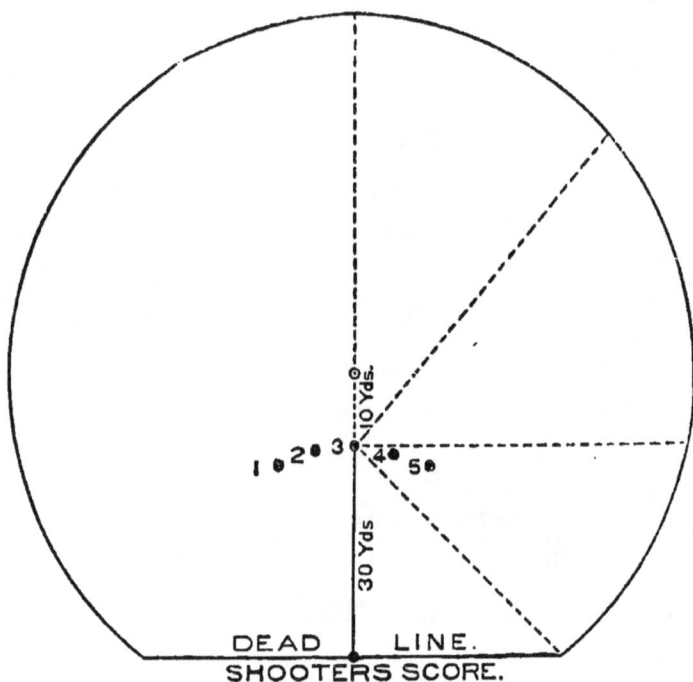

Diagram D.   (See Rules 5 and 7—Live Bird Shooting.)

NOTE.—This should give from center trap to boundary, to straightaway bird,
60 yards; to right quarterer, 58 yards; to bird at right angles, 48 yards; to junction
of circle and dead line, 42 yards.

RULE 11. *Mutilating Birds.*—No mutilation of birds will be allowed, and if it is proved to the referee that any contestant has wilfully mutilated a bird, or is a party thereto, the referee shall declare all his rights in the match forfeited.

RULE 12. *Out of Bounds.*—A bird once out of bounds must be scored a "Lost bird."

RULE 13. *Birds Shot at by Another Person.*—If a bird be shot at by any person other than the shooter at the score, the referee shall decide whether it shall be scored, or another bird allowed.

RULE 14. *Position of Gun.*—Any the shooter may adopt.

RULE 15. *Loads.*—Charges of powder unlimited. Charge of shot not to exceed one and one-quarter ounces American Association, or Dixon's measure, struck. Any shooter using a larger quantity of shot shall forfeit his entrance money and rights in the match.

RULE 16. *Caliber and Weight of Gun.*—No gun of larger caliber than a 10-bore shall be used, and the weight of all guns shall be unlimited.

RULE 17. *Loading Guns.*—No gun shall be loaded except at the score. Cartridges must be removed from the gun before leaving the score.

RULE 18. · *Gun not Cocked.*—If a gun is not cocked, or the safety not properly adjusted, and the bird escapes, it shall be scored a "Lost bird."

RULE 19. *Miss-Fire with the First Barrel.*—If the shooter's gun miss-fire with the first barrel, and he uses the second barrel and misses, the bird must be scored "Lost bird." But if killed with the second barrel, on the wing, it shall be scored "Dead bird."

RULE 20. *Miss-Fire with the Second Barrel.*—If a miss-fire occur with the second barrel, the shooter shall have another bird, using a full charge of powder only in the first barrel. He must, however, put the gun to his shoulder and discharge the blank cartridge in the direction of the bird, and the bird must be on the wing when the first barrel is discharged.

RULE 21. *Shooter at the Score.*—In all contests the shooter must be at the score within three minutes after his name is called to shoot, or he forfeits his rights in the match.

RULE 22. *Leaving the Score.*—A shooter having fired his first barrel and left the score, cannot return to fire his second barrel.

RULE 23. *Balk.*—If a contestant is balked or interfered with, or there is other similar reason why it should be done, the referee may allow another bird.

RULE 24. *Announcing the Score.*—The referee shall announce the result of each shot distinctly and it shall be called back by the scorer, and at the close of each shooter's score the result must be announced, and if claimed to be wrong, the error, if any, must be corrected at once.

RULE 25. *Tie Shooting.*—All ties shall be shot off at the original distance and as soon after the match as practicable, at the following number of birds:
   In matches of 10 birds or less, 3 birds.
   In matches of 11 to 25 birds, inclusive, 5 birds.
   In matches of 26 to 50 birds, inclusive, 10 birds.
   In matches of 51 to 100 birds, inclusive, 25 birds.
   If in a series of matches the result prove a tie, such tie shall be shot off at the original number of birds.

RULE 26. *Class Shooting.*—All shooting shall be class-shooting, unless · otherwise stated.

RULE 27. *Endangering Person or Property.*—If a bird shall fly so that to shoot at it would endanger any person or property, it shall not be shot at, and the referee shall allow another bird.

RULE 28. *Forbidden Shooting.*—No shooting shall be permitted within the enclosure other than at the score, and in case there is no enclosure, no shooting within 200 yards of the score, except by those at the score, without the consent of the management.

DOUBLE BIRDS.

RULE 1.   The rules for single bird shooting shall govern double bird contests, when not conflicting with the following:

RULE 2.   *Double Rises.*—The double rises shall be from two traps of any kind, 10 yards apart, pulled simultaneously.   The rise shall be:
    26 yards for 10-bore guns.
    24 yards for 12-bore guns.
    22 yards for 14 and 16-bore guns.
    21 yards for 20-bore guns.

RULE 3.   *Allowing Another Pair.*—Both birds should be on the wing when shot at.   Should only one bird fly, the shooter shall have another pair of birds if he does not shoot, or if he does shoot and kills the bird on the wing.   But if he shoots and misses, the bird shall be scored lost, and in such event he shall shoot at another pair of birds, with a full charge of powder only in one barrel.   The referee shall load the gun, not allowing the shooter to know which barrel contains the full charge, and which contains the powder charge only.

RULE 4.   *Miss-Fire with the First Barrel.*—If the shooter's gun miss-fire with the first barrel, he will be entitled to another pair of birds if he does not shoot his second barrel.   But if he fires his second barrel, the result must be scored, and the shooter shall shoot at another pair of birds, with a full charge of powder only, in one barrel, as provided for in Rule 3.

RULE 5.   *Miss-Fire with the Second Barrel.*—If the shooter's gun miss-fire with the second barrel, the result of the first barrel must be scored, and the shooter shall shoot at another pair of birds with a full charge of powder only, in one barrel, as provided for in Rule 3.

RULE 6.   *Lost Bird.*—   If a shooter fire both barrels at one bird intentionally, it shall be scored "Lost birds."   But if the second barrel be discharged accidentally, it shall be "No birds."

RULE 7.   *No Bird.*—If both pair are killed with one barrel, it shall be declared "No birds," and the shooter shall shoot at another pair of birds.

RULE 8.   *Ties.*—All ties must be decided in shooting off as follows:
    In matches of 5 pairs or less, at 2 pairs.
    In matches of 6 to 10 pairs, inclusive, at 3 pairs.
    In matches of 11 to 20 pairs, inclusive, at 6 pairs.
    In matches of 21 to 50 pairs, inclusive, at 10 pairs.

LONDON GUN CLUB RULES.

1.   A miss-fire with first barrel is no shot under any circumstances.   If the shooter miss-fire with the second barrel he shall have another shot, but with the ordinary charge of powder, and no shot in the first barrel.

2.   If the gun be locked, or not cocked, or not loaded, and the bird flies away, it is a "lost bird;" if the stock or cock should break in the act of firing it is "no bird."

3.   If the trap is pulled without notice from the shooter, he has the option to take the bird or not.

4.   The puller shall not pull the trap until the trapper and the dog are back in their places, even should the shooter call "pull."

5.   If, on the trap being pulled, the bird does not rise, the shooter to take it or not at his option; but if not, he must declare it by saying "no bird," before it is on the wing.   If, however, the bird rises and settles before the shooter fires, it shall be at his option to refuse it or not.

6.   *Single Shooting.*—If more than one bird be liberated, it is "no bird."

7.   In shooting at a bird, should both barrels go off at once it shall score the same as if they had been let off "separately."

8.   *Double Shooting.*—If more than two traps be pulled they are "no birds;" should both birds not rise simultaneously on the opening of the traps, they are "no birds."

9. A bird to be scored good must be gathered by the dog or man without the aid of a ladder or any other instrument, and all birds not gathered in the ground, or gathered inside the pavilion enclosure, having flown over the railings, to be scored lost.

10. If a bird which has been shot perches or settles on the top of the fence, or any of the buildings in the ground higher than the fence, it is to be scored a "lost bird."

11. If a bird once out of the ground return and fall dead within the boundary, it must be scored a "lost bird."

12. If the first barrel be fired whilst the bird is on the ground, should the bird be killed by either barrel it is "no bird;" if missed, it is lost. It may be shot on the ground with the second barrel if it has been fired at with the first barrel while on the wing.

13. The shooter is bound at once to gather his bird, or depute some person so to do when called upon; but in so doing he must not be assisted by any other person or use any description of implement. Should the shooter be in any way baffled by his opponent, or by any other person or dog, he can claim another bird with the sanction of the referee.

14. The shooter having once left the mark after shooting at the bird, cannot shoot at it again under any circumstances.

15. In matches or in sweepstakes any shooter found to have in his gun any more shot or powder than is allowed, to be at once disqualified.

16. Any shooter is compelled to unload his gun on being challenged; but if the charge is found not to exceed the allowance, the challenger shall pay £1 to the shooter, which must be paid before he (the challenger) shoots again.

17. Officers of the Army and Navy on full pay, provided they are *bona fide* guests of a member for the day, are allowed to shoot in any sweepstakes to which the club does not add prize or money. Members accredited from foreign clubs shall only be allowed to shoot for four weeks during the season, after which they must be proposed and seconded if they desire to shoot.

18. Breech-loaders not to be loaded until the shooter is at the mark, and the trapper has returned to his place. On leaving the mark, should a cartridge not have been discharged, it is to be removed before the shooter turns his face from the traps.

19. No wire cartridges allowed; nor is any bone dust or other substance to be mixed with the shot.

20. Should any shooter shoot at a distance nearer than his proper distance, the bird, if killed, is "no bird," if missed, a "lost bird;" or should he, by direction of the referee or scorer, shoot at any wrong distance, the bird, if missed, shall be "no bird," and the shooter shall be allowed another, which, if killed, shall be scored.

21. 1¼ oz. of shot and 4 drachms of black powder, or its equivalent in any other description of gunpowder, is the maximum charge. In advertised handicaps the shooter is allowed to go in half a yard for every ⅛ of an ounce of shot less than the maximum.

22. In shooting for the principal advertised events members can enter up to the end of the second round, unless it shall be within the knowledge of the referee that any member proposing to enter has been on the ground during the first round, in which case he shall not be permitted to shoot after the commencement of the second round; for all other sweepstakes entries must be made before the end of the first round; special sweepstakes excepted.

23. The sweepstakes preceding the chief event of the day shall be divided by those shooters who may be in at the end of the round at three o'clock in equal proportions.

24. That the baskets containing the birds for the whole day's shooting be numbered by paint at the back. That the baskets, in the order they are to be brought out and trapped, be drawn for by lot by the referee, and that the baskets so marked be used in the order of rotation in which they are drawn.

REVISED GAME LAWS
FOR
Western States and Territories.

GAME AND FISH LAWS OF ARIZONA.

An act for the preservation of the Game Animals, Birds and Fishes of Arizona, and amendatory to part I, Title 16, Penal Code Revised Statutes of Arizona.

§ 1. Any person who in this Territory shall, between the first day of January and the first day of September in each year, hunt, pursue, take, kill or destroy any male or female deer, or fawn, or antelope, is guilty of a misdemeanor.

§ 2. Any person who shall in this Territory, at any time during the period of five years next subsequent to the passage of this Act, hunt, pursue, take, kill or destroy any elk, male or female, or any mountain sheep or big horn, male or female, is guilty of a misdemeanor.

§ 3. Any person who shall in this Territory, between the first day of April and the first day of September in each year, shoot, kill, trap or destroy any wild turkey, grouse, prairie chicken, partridge or quail, is guilty of a misdemeanor. The provisions of this paragraph are not to be construed as applying to individuals killing quail on their own premises inclosed.

§ 4. Every person who shall be found hunting with gun or dog, within the enclosure of another, and who shall take, kill or destroy any of the game mentioned in the foregoing paragraphs of this Act, without the consent of said owner or authorized agent, is guilty of a misdemeanor, and upon conviction, shall be fined or imprisoned, as hereinafter prescribed.

§ 5. Any person who shall in this Territory, between the first day of November and the first day of April in each year, take, catch or procure any of the brook or mountain trout, or who shall at any time sell or offer for sale any of the brook or mountain trout procured from any of the streams and lakes of this Territory, is guilty of a misdemeanor. The provisions of this paragraph are not to be construed as applying to rights of individuals over private lakes, ponds or pools.

§ 6. The possession of the dead bodies of any of the game hereinbefore mentioned, when the taking and killing of the same is prohibited, is a misdemeanor.

§ 7. If any agent or servant of any railroad company, express company or other common carrier or private individual, have or receive for transportation or carriage, or for any other purpose, any of the game, animals, birds or fishes mentioned in this Act, when the killing and taking of same is prohibited, or for transportation or carriage outside of the limits of this Territory, at any time they shall be guilty of a misdemeanor.

§ 8. It shall be unlawful for any person or persons to take, kill or destroy any fish with giant powder or any other explosive, or in any other manner except with hook and line or spear. Any violations of the provisions of this paragraph shall be deemed a misdemeanor.

§ 9. Any person found guilty of violation of any of the provisions of the various paragraphs of this Act, shall, upon conviction, be fined in the sum of not less than fifty dollars ($50.00) or more than one hundred dollars ($100.00) and costs, and in default of payment of such fine and costs, be imprisoned in the county jail for a period not to exceed one day for each dollar of such fine and costs unpaid. It is hereby made the duty of any peace officer to arrest any person or persons violating any of the provisions of this Act, and take them before any civil authority competent to try the offense.

§ 10. All Acts embodied in paragraphs 993, 994, 995, 996, 997, 998, 999 and 1000, page 752, Revised Statutes of Arizona, and not embodied in this Act, are hereby repealed.

§ 11. This Act to take effect and be in force from and after its passage.

Approved April 12, 1893.

CALIFORNIA.

PENAL CODE AS AMENDED IN 1895.

§ 626. **Valley Quail, Bob White, Partridge, Robin, Wild Duck, Rail.**—Every person who, between the 15th day of February and the 15th day of October in each year, shall kill or have in his possession, whether taken in the State of California or shipped into the state from any other state, territory or foreign country, except for purposes of propagation, any valley quail, bob white, partridge, robin, or any kind of wild duck or rail, shall be guilty of a misdemeanor.

§ 626A. **Mountain Quail, Grouse.**—Every person who, between the 15th day of February and the 15th day August in each year, shall kill or have in his possession, whether taken or killed in the State of California or shipped into the state from any other state, territory or foreign country, except for purposes of propagation, any mountain quail or grouse, shall be guilty of a misdemeanor.

§ 626B. **Doves.**—Every person who, between the 15th day of February and the 1st day of July in each year, shall kill or have in his possession any dove or doves, shall be guilty of a misdemeanor.

§ 626C. **Male Deer.**—Every person who shall kill any male deer between the 15th day of October and the 15th day of July of the following year shall be guilty of a misdemeanor.

§ 626D. **Female Deer, Fawn, Antelope, Elk, Mountain Sheep.**—Every person who shall at any time kill any female deer, or spotted fawn, or any antelope, elk or mountain sheep, shall be guilty of a misdemeanor.

§ 626E. **Hides and Venison.**—Every person who shall at any time buy, sell or offer for sale the hide or meat of any deer, elk, antelope or mountain sheep, whether taken or killed in the State of California or shipped into the state from any other state or territory, shall be guilty of a misdemeanor; provided that nothing in this section shall be held to apply to the hide of any of said animals taken or killed in Alaska or any foreign country.

§ 6:6F. Every person who shall buy, sell, offer or expose for sale, transport or carry or have in his possession the skin, hide or pelt of any deer from which the evidence of sex has been removed, shall be guilty of a misdemeanor.

§ 626G. **Pheasant.**—Every person who shall, within the three years next after the passage of this act, kill or have in his possession, except for the purposes of propagation, any pheasant, shall be guilty of a misdemeanor.

§ 627. **Bore of Gun.**—Every person who shall use a shot gun of a larger caliber than that commonly known and designated as a No. 10 gauge shall be guilty of a misdemeanor. The proof of the possession of said gun in the field, on marsh, bay, lake or stream, shall be *prima* facie evidence of its illegal use.

§ 627A. **Trespass.**—Every person who upon any inclosed or cultivated grounds which are private property, and where signs are displayed forbidding such shooting, except salt-water marsh land, shall shoot any quail, bob white, pheasant, partridge, grouse, dove, deer or wild duck, without permission first obtained from the owner or person in possession of such grounds, or who shall maliciously tear down, mutilate or destroy any sign, signboard or other notice forbidding shooting on private property, shall be guilty of a misdemeanor.

§ 627B. **Exportation.**—Every railroad company, express company, transportation company or other common carrier, their officers, agents and servants, and every other person who shall transport, carry or take out of this state, or shall receive for the purpose of transporting from the state, any deer, deer skin, buck, doe or fawn, or any quail, partridge, pheasant, grouse, prairie chicken, dove or wild duck, except for purposes of propagation, or who shall transport, carry or take from the state, or receive for the purpose of transporting from this state, any such animal or bird, shall be guilty of a misdemeanor.

§ 627C. **Song Birds.**—Every person who shall at any time kill or have in his possession, except for the purpose of propagation, or for educational or scientific purposes, any English sky lark, canary, California oriole, humming bird, thrush or mocking bird, or any part of the skin, skins or plumage thereof, or who shall rob the nests, or take or destroy the eggs of any of the said birds, shall be guilty of a misdemeanor.

§ 628. **Striped Bass, Black Bass.**—Every person who takes or has in his possession any striped bass of less than three pounds in weight is guilty of a misdemeanor. Every person who, between the 1st day of January and the 1st day of July, takes or catches, buys, sells or has in his possession, any black bass, is guilty of a misdemeanor.

§ 632. **Trout.**—Every person who at any time takes or catches any trout, except with hook and line, is guilty of a misdemeanor. § 633. Every person who takes or has in his possession any speckled trout, brook or salmon trout, or any variety of trout, between the 1st day of November and the 1st day of April in the following year, is guilty of a misdemeanor; provided, however, that steelhead trout may be possessed at any time when taken with rod and line in tide water. Every person who buys or sells, offers or exposes for sale, within this state, any kind of trout less than six inches in length, is guilty of a misdemeanor.

§ 634. **Salmon.**—Every person who between the 31st day of August and the 1st day of November of each year, takes or has in his possession any fresh salmon, is guilty of a misdemeanor.

Supervisors' Ordinances.—Supervisors are empowered to provide by ordinance, not in conflict with the general laws of the state, for the protection of fish and game, and may shorten the close season for the taking or killing of fish and game within the dates fixed by the general state laws, but shall not lengthen them.

COLORADO.

§ 1. Act of April 7, 1893. **Insectivorous Birds, Mongolian Pheasant, Quail, Ptarmigan, Partridge, Dove.**—No person shall kill any robin, lark, whip-poor-will, finch, sparrow, thrush, wren, martin, swallow, snowbird, bobolink, red-winged blackbird, crow, raven, turkey buzzard, oriole, kingbird, mockingbird, song sparrow, or other insectivorous bird, or any mongolian pheasant, quail, ptarmigan, partridge or dove, at any time; provided, that doves may be shot from July 15 to Oct. 1.

§ 2. **Wild Turkey, Pheasant, Prairie Chicken, Grouse.**—No person shall kill any wild turkey, pheasant, prairie hen, prairie chicken, or grouse, except that they may be shot between Aug. 15 and Nov. 1 of the same year.

§ 3. **Wild Fowl.**—No person shall kill wild ducks, geese, brant or swans, or other water fowls, except between the 1st day of September and the 1st day of May following, during which time the same may be killed by means of an ordinary shoulder gun; and it shall be unlawful to use any swivel or punt gun for said purposes. § 5. No person shall during the night time, by the use of any artificial light, or any like device whatever, kill any wild duck, wild goose, brant, swan, or other water fowl or fish.

§ 8. **Dogs for Big Game.**—It shall be unlawful for any person to use dogs for the purpose of running or coursing mountain sheep, deer, elk or antelope.

§ 11. **Bison, Buffalo, Mountain Sheep.**—No person· shall kill at any time any bison, buffalo or mountain sheep.

§ 12. **Deer, Elk, Antelope.**—No person shall kill any deer, elk or antelope at any time, except that those deer, antelope and elk which have horns may be killed between August 1st and November 1st of the same year, for good [*sic*] purposes only, and then only when necessary for immediate use, governed in amount and quantity by the reasonable necessity of the person or persons killing the same. "Reasonable necessity" shall be construed to mean not more than one elk, deer or antelope in the possession of any one person at one time.

General Statutes, Chap. 40 [as amended 1893]. **Trout, Food Fish.**—That it shall not be lawful to take any trout or other food fish during the months of December, January, February, March, April and May, or either of said months; it shall be unlawful to kill any trout, or other food fish during the open season, except for food, and then only when necessary for immediate use, governed in amount and quantity by the reasonable necessities of the person or persons catching such fish. It shall be unlawful to sell any trout or other food fish, or to ship or transport them out of this State. It shall be unlawful to kill any trout less than six inches in length.

IDAHO.

§ 1. Act March 11, 1895. **Moose, Caribou, Mountain Sheep, Mountain Goat, Elk.**—It shall be unlawful to kill any moose, caribou, mountain sheep, mountain goat or elk at any time before the 1st day of September, 1897, and after said 1st day of September, 1897, at any time except between the 1st day of September and the 31st day of December of each year.

§ 2. **Deer, Antelope.**—It shall be unlawful to kill any deer or antelope except between the 1st day of September and the 31st day of December of each year: or to kill, ensnare or trap any of the animals mentioned in this and the preceding section at any time of the year solely for the purpose of buying or selling their carcasses. Provided that any of the animals mentioned in this and the preceding section may be taken alive, which shall be for scientific purposes only.

§ 4. **Dogs Forbidden.**—It shall be unlawful to hunt or chase with dogs any moose, caribou, elk, deer, antelope, mountain sheep or mountain goat; or to own, keep or have in possession any dog or dogs for said purpose.

§ 6. **Mongolian Pheasant.**—It shall be unlawful to kill any Mongolian pheasant at any time before the 1st of August, 1900, and after said 1st day of August, 1900, at any time except between the 1st day of August and the 15th day of December of each year.

§ 7. **Quail.**—It shall be unlawful to kill any quail except between the 1st day of October and the 1st day of December of each year.

§ 8. **Partridge, Pheasant, Grouse, Prairie Chicken, Sage Hen, Fool Hen.**—It shall be unlawful to kill any partridge, pheasant, grouse, prairie chicken, sage hen or fool hen, except between the 1st day of August and the 15th day of December of each year.

§ 9. **Wildfowl.**—It shall be unlawful to kill any species of wild duck, goose or swan, or to take or destroy their eggs, between the 15th day of April and the 15th day of August of each year.

§ 12. **Fish.**—It shall be unlawful to catch or kill any species of fish, except salmon, salmon trout and sturgeon, in any of the streams, rivers, lakes, reservoirs or waters lying within the State of Idaho, with any seine or other net, or any spear, weir, fence, basket, trap, gill net, set net, otter or any other contrivance, except hook and line attached to a pole or held in the hand. Provided that no fish shall be taken by any method, except for breeding (and home consumption), between the 1st day of November and the 15th day of May of the succeeding year.

KANSAS GAME LAWS.

CHAPTER 97.

An act for the protection of birds, and naming what birds shall not be killed, and prescribing punishment for the violation of the provisions of this act, and to repeal paragraphs 3195, 3196, 3198, 3199, 3200, 3201, 3202 and 3203 of the general Statutes of 1889.

Be it enacted by the Legislature of the State of Kansas.

§ 1. **Birds not to be killed.**—It shall be unlawful for any person or persons, at any time, to catch, kill, shoot, trap, or ensnare, any partridge, prairie chicken, grouse, quail, pheasant, oriole, meadow lark, redbird, mocking bird and bluebird. Provided, that no provisions of this act shall apply or interfere with persons who may have in their possession or raise for sale any birds as pets, or may at any time, catch, kill, or entrap any of the birds mentioned in this section on his or her own premises, controlled by such person for his or her own use.

§ 2. **Cannot buy or sell.**—It shall be unlawful for any person, company or corporation, at any time, to buy, sell, or barter within the State of Kansas, any bird or birds named in section one of this act, except the song birds mentioned in section one of this act. The having in possession, by any person, company or corporation, of any birds named in section one of this act, except the song birds mentioned in section one of this act, shall be deemed prima facie evidence of the violation of this act.

§ 3. **Penalty.**—Any person, company or corporation found guilty of violation of any of the provisions of this act, shall be deemed guilty of a misdemeanor, and upon conviction thereof before any court of competent jurisdiction, shall be fined in a sum not to exceed twenty-five dollars, for each and every offense, and costs, together with attorney's fee, of ten dollars, and shall be committed until such fine, costs and attorney's fee shall be paid.

§ 5. It shall not be necessary to prove on the trial, or to state in the complaint the true or ornithological name of the bird caught, killed, shot, trapped, ensnared or had in possession in violation of this act.

§ 6. **Specimens for Scientific Purposes.**—The provisions of this act shall not apply to any person who shall catch or kill any wild bird or birds, for the sole purpose of preserving them as specimens for scientific purposes. Provided, that in a prosecution for a violation of any of the provisions of this act, it shall not be deemed necessary for the prosecution to set up or prove that the killing, catching or having in possession of any wild bird was not done for scientific purposes.

§ 7. That chapter forty-five (45), of the General Statutes of 1889, relating to game, being paragraphs 3195, 3196, 3198, 3199, 3200, 3201, 3202 and 3203 thereof, and all acts and parts of acts inconsistent herewith, be and the same are hereby repealed.

§ 8. This act shall take effect on and after its publication in the Statute book.

Approved March 11, 1893.

§ 4. Act in effect April 5, 1895. **Black Bass.**—It shall be unlawful for any person to capture any black bass between the 1st day of April and the 1st day of July; or at any time to take any black bass of less than eight inches in length.

MISSOURI.

§ 1. Chap. —, Laws of 1895. **Deer, Birds**—It is hereby declared unlawful to kill any deer in the State of Missouri under one year of age. It is further declared unlawful to kill any deer of any age between the first day of January and the first day of October in each year; and for the purpose of preventing the extinction of the species it is hereby declared unlawful to kill any doe within five years after the passage of this Act. It is further declared unlawful to kill any wild song bird or insectivorous bird at any season of the year, or to disturb, rob or destroy the nests of such birds, or take therefrom any egg or eggs. It is further declared unlawful to kill any wild turkey, pinnated grouse (commonly called prairie chicken), or ruffled grouse (commonly called pheasant or partridge), or any quail (sometimes called Virginia partridge), between January 1 and November 1 of each year, or any woodcock, turtle dove, meadow lark or plover, between January 1 and August 1 in each year. And it is further declared unlawful at any time or season to catch, take or injure by means of nets, traps, pens or pits or other device, any kind of game as herein described, within this State; and every person who shall kill any wild duck between the first day of April and the first day of October, or who shall ensnare, trap or kill by means of any explosive any wild goose or duck, or who shall shoot or kill the same between sunset and sunrise by means of gunpowder or other explosives of any kind, shall be guilty of a misdemeanor. Provided, that the provisions of this section shall not apply to any person who shall ensnare, trap or net wild geese or ducks on his own premises for his own use.

§ 3905. Rev. Stats. **Non-Residents**—If any person being a non-resident of this State, shall kill any deer, fawn, wild turkey, pinnated grouse, ruffled grouse, quail, woodcock, goose, brant, duck or snipe, coon, mink, otter, beaver, bear, muskrat or other furred animals, he shall be deemed guilty of a misdemeanor.

§ 1. Act of April 8, 1895. **Shipment of Game**—It shall be unlawful for and during the period of five years next succeeding the passage of this act, for any person to ship from the county where killed to any other county in the State or to any point outside of the State, any quail, pinnated grouse, prairie chickens, deer or turkey.

Fish—No restrictions on angling.

Imported Game—Act March 11, 1895, protects California valley or mountain quail or partridge, Mexican or Texas quail or partridge, of any variety of Messena or Montezuma to 1900.

MONTANA.

[CAUTION.—We are advised that the act adopting the Code of 1895 did not repeal the act of March 1, 1893. The conflicting provisions of the two laws (and conflicting § 701 and 702 of the Code) protect game during the following periods: Bison, buffalo, quail, Chinese pheasant, to 1903. Moose, elk, otter, beaver, to 1899. Deer, goat, sheep, antelope, Dec. 15 to Sept. 15 following. Grouse, prairie chicken, pheasant, fool-hen, sage hen, partridge, snipe, Nov. 15 to Aug. 15 following. Wildfowl, Jan. 1 to Sept. 1. Speckled trout, May and June.]

ACT APPROVED MARCH 1, 1893.

§ 1. **Bison, Buffalo, Quail, Chinese Pheasant, Moose, Elk, Otter.**—That any person who shall kill for the period of ten years from and after the passage of this act, any bison, buffalo or quail or Chinese pheasant, or who shall kill for the period of six years from and after the passage of this act, any moose, elk, otter or beaver, shall be deemed guilty of a misdemeanor.

§ 2. **Deer, Sheep, Goat, Antelope.**—That any person who shall kill any white-tail deer, black-tail deer, mule deer, mountain sheep, Rocky

Mountain goat and antelope, between the 15th day of December and the 15th day of August of the following year, shall be deemed guilty of a misdemeanor. § 6. That any person who shall kill at any time any of the animals mentioned in § 2, for the purpose of procuring the head or hide only, or for speculative purposes, or for market or for sale, shall be deemed guilty of a misdemeanor.

§ 9. **Grouse, Prairie Chicken, Pheasant, Fool-Hen, Sage Hen, Partridge, Snipe.**—That any person who shall kill any grouse, prairie chicken, pheasant, fool-hen, sage hen, partridge or snipe, between the 15th day of November and the 15th day of August of the next ensuing year, shall be deemed guilty of a misdemeanor.

§ 10. **Wildfowl.**—That any person who shall kill any wild geese or wild ducks, brant or swan, between the 1st day of January and the 1st day of September of each year, shall be deemed guilty of a misdemeanor.

§ 4. **Fish.**—That a fishing tackle consisting of a rod or pole, line and hooks, shall be the only lawful means by which fish may be taken from any of the waters of the State, provided, that no speckled mountain trout, or other small fish, shall be taken, except from private ponds by owners, in any manner during the months of May and June.

§ 1. **Act of March 12, 1885. Exportation of Skins.**—It shall be unlawful to ship, carry or cause to be carried in any manner whatsoever, from the Territory of Montana to any other Territory or State, the skin of any moose, deer, elk, bison, buffalo, antelope or mountain sheep; provided, that nothing in the provisions of this act shall prevent the shipment of any specimens that are stuffed or mounted as curiosities.

PENAL CODE, PT. I., CHAP. I., TITLE XV., APPROVED FEB. 19, 1895.

§ 700. **Buffalo, Elk, Otter, Beaver, Quail, Chinese Pheasant, Moose.**—Every person who kills any bison, buffalo, quail or Chinese pheasant, or any female moose, female elk, otter or beaver; or who, between December 15th of one year and September 1st of the following year, kills any male moose or elk; or who, in a single open season, shall kill more than two bull moose or three bull elk, or one hundred grouse or prairie chickens, is punishable by a fine.

§ 701. **Deer, Mountain Sheep, Rocky Mountain Goat, Antelope.**—Every person who kills any deer, mountain sheep, Rocky Mountain goat or antelope between the 1st day of January and the 15th day of September of the same year is punishable by a fine.

§ 702. **Deer, Mountain Sheep, Mountain Goat, Antelope.**—Every person who, between January 1st and September 1st of each year, kills any deer, mountain sheep, Rocky Mountain goat or antelope, or who in any single calendar year's open season kills more than eight deer, eight mountain sheep, eight Rocky Mountain goats, or eight antelope, is punishable by a fine.

§ 703. **Use of Dogs.**—Every person who wilfully hunts, chases or runs with dogs any of the animals mentioned in the three preceding sections at any time, is punishable by fine. The use of dogs, however, to capture any such animals after they have been wounded is not a violation of this section.

§ 704. **Prairie Chicken, Fool-Hen, Grouse, Pheasant, Partridge.**—Every person who, between the 1st day of January and the 15th of August of each year, kills any grouse, prairie chicken, fool-hen, pheasant or partridge; or who at any season of the year shall, for speculation, market or sale, kill any of the above mentioned birds, is punishable by a fine. [For lawful number see § 700.]

§ 705. **Wildfowl.**—Every person who kills any wild geese or wild ducks between the 1st day of May and the 15th day of August of the same year is punishable by a fine.

§ 706. **Other Birds.**—Every person who kills any robin, meadow-lark, bluebird, thrush, oriole, woodpecker, mockingbird, goldfinch, snowbird, cedarbird, or any other of the small birds known as singing birds, is punishable by a fine,

NEBRASKA.

§ 5666. Compiled Statutes. **Wildfowl.**—It shall be unlawful at any time, by the use of any swivel, punt gun, big gun, or any gun other than the common shoulder gun, or any punt-boat or sneak-boat used for carrying such gun, to kill any wild goose, wood duck, teal, canvas-back, blue-bill, or other wild duck. [No close season.]

§ 5667. **Buffalo, Elk, Mountain Sheep, Deer, Antelope, Grouse, Quail, Wild Turkey.**—It shall be unlawful for any person to kill any wild buffalo, elk, mountain sheep, deer or antelope, between the 1st day of January and the 1st day of October; or to kill any grouse between the 1st day of January and the 1st day of September; or to kill any quail or wild turkey between the 1st day of January and the 1st day of October. It shall be unlawful for any person to transport, or ship any such animal or bird at any time of the year. Transportation of game prohibited at all times of the year. Unlawful to kill any Mongolian, Japanese, copper, trogopan, silver or golden pheasants for six years from March 30, 1895. Hounding of deer prohibited in Burt, Washington, Douglas, Sarpy, Cass, Saunders and Dodge counties.

Fish.—Fish can be taken by hook and line only. Fish planted by the Fish Commission or private persons protected at all times.

NEVADA.

§ 1. Chap. 49, Laws 1893. **Song Birds.**—It shall be unlawful to kill any sparrow, bluejay, martin, thrush, mocking-bird, redbreast, cat-bird, wren, robin, meadow-lark, or humming-bird; or any song bird, except linnets.

§ 2 (as amended 1895). **Game Birds.**—It shall be unlawful, at any time between the 1st day of April of any year and before the 15th day of September following, to kill any wild goose, wood-duck, teal, mallard or other ducks, sand-hill crane, brant, swan, plover, curlew, snipe and mud hens; and between the 15th day of March and 15th day of September any partridge, pheasant, woodcock, grouse, quail, bittern or yellow-hammer.

§ 3 (as amended 1895). **Sage Cock.**—It shall be unlawful to kill, between the 1st day of March of every year and before the 15th day of July following, any sage cock, hen or chicken.

§ 4 (as amended 1895). **Big Game.**—It shall be unlawful to kill at any time after the 1st day of January and before the 1st day of September of each year, any deer, antelope, elk, mountain sheep, goat or caribou.

§ 5. **Beaver, Otter.**—It shall be unlawful to kill any beaver or otter before the 1st day of April, 1897.

§ 6. **Scientific Purposes.**—Provided, that nothing in this act shall be so construed as to prohibit any person taking any bird, fowl or animal mentioned in this act, at any time, for scientific purposes.

§ 2 (as amended 1895). Chap. 72, Laws 1891. **Trout, Land-locked Salmon.**—It shall not be lawful, between the first day of October of each year and the 1st day of June of each year, to catch or kill any river, lake or brook trout, or land-locked salmon, in any of the streams, rivers, lakes or other waters within this State; provided, that the "close season" as to lake trout in all the lakes of this State shall commence on the 1st day of January of each year and end on the 1st day of May of each year; and provided further, that the close season as to the waters of the Humboldt River and its tributaries shall commence on the 1st day of November of each year and end on the 1st day of June of each year.

NEW MEXICO.

§ 4. Chap. 53, Laws 1889. **Trout, Food Fish.**—That it shall not be lawful to kill any trout or other food fish during the months of November, December, January, February, March and April.

§ 5. For Food Only.—It shall be unlawful to kill any trout or food fish except for food, and then only when necessary for immediate use, governed in amount and quantity by the reasonable necessities of the person or persons catching such fish.

§ 1. Act of Feb. 16, 1895. Deer, Elk, Antelope—No person shall kill any deer, elk, fawn or antelope between the 1st day of January and the 1st day of October.

§ 2. Wild Turkey, Quail—No person shall kill, ensnare or trap any wild turkey or quail between the 1st day of March and the 1st day of October; provided, further, that it shall be unlawful to ensnare or trap quail at any time.

§ 4. Wanton Killing, or for Skins—No person shall wantonly kill any of the game, birds or animals mentioned in this Act; nor shall it be lawful at any time to kill any elk, deer, fawn or antelope for the sole purpose of securing the hide or skin of any such animal.

§ 5. Export—That it shall be unlawful for any railway, express company or any common carrier, or their agents or employees, to transport outside of the territory, or receive for such transportation, at any time, any of the flesh or meat of any of the birds or animals named in this Act that may be offered for transportation at any station or place in this territory.

NORTH DAKOTA.

Act of April —, 1895. Non-Residents.—It shall be unlawful for any non-resident of this State to hunt, kill or wound in this State any of the wild animals or birds mentioned in the Penal Code, without having first obtained the permit for non-residents hereinafter provided for. It shall be unlawful for any resident of this State, either alone or with another or others, to use any dog in hunting, setting, pointing or retrieving any game unless such resident shall hold a permit good at the place where he so uses a dog, as hereinafter provided. Such permits shall be subject to inspection of any person upon demand. [License fee for non-residents, $25; for residents, 50 cents; license good for one season].

§ 861. Game Seasons, Methods, Amount.—Every person who either (1) shoots or kills any prairie chicken, pinnated grouse, sharp-tailed grouse, ruffed grouse, woodcock, plover, wild duck, wild goose or brant, between the 1st day of December and the first day of September following, or any song bird at any time; or

(2) At any time kills or shoots any wild duck, goose or brant with any swivel gun or other gun, except such as is commonly shot from the shoulder, or in hunting such birds makes use of any artificial light or battery; or

(3) Uses or employs any trap, snare, net or bird lime, or medicated, drugged or poisoned grain or food, to capture or kill any of the birds mentioned in Subdivision 1 of this section; or

(4) Wantonly destroys any nest of eggs of any of the birds mentioned in Subdivision 1 of this section; or

(5) Shoots or kills any buffalo, elk, deer, antelope, caribou or mountain sheep between the 15th day of December and the 1st day of November following; or

(6) At any time uses or employs any hound or dog in running or driving any of the animals mentioned in Subdivision 5 of this section; or

(7) Sets any spring or other gun, trap, snare or any other device to kill, wound or destroy any of the animals mentioned in Subdivision 5 of this section; or

(8) Knowingly hunts in any way upon the enclosed or cultivated lands of another without the consent of the owner, or his agent or tenant; or

(9) Shoots or kills in any one day more than twenty-five of the game birds mentioned in Subdivision 1, or more than five of the animals mentioned in Subdivision 5 of this section, is guilty of a misdemeanor.

§ 862A. Fish.—Every person who either (1) takes, catches, kills or destroys any fish of any kind in any of the lakes, streams or other waters of this State, except the Missouri River, in any manner other than by angling with hook and line; or

(2) Between the 1st day of November and the following 1st day of May in each year takes, catches kills or destroys in any manner or by any device, in any of the waters of this State, except the Missouri, any pike, pickerel, perch, bass or muscalonge, for any purpose other than propagating or breeding same; or

(3) Exposes any such fish for sale during such period, is guilty of a misdemeanor.

Export.—Every person who within this State ships or receives for shipment beyond the limits of this State any of the game birds or animals mentioned in Section 861, or any of the fish mentioned in Section 862A, is guilty of a misdemeanor.

OREGON.

ACT OF FEB. 28, 1895.

§ 1. **Elk, Moose, Mountain Sheep, Deer.**—Every person who shall at any time between the 1st day of December of each year and the 1st day of August of the following year kill any elk, moose or mountain sheep, shall be guilty of a misdemeanor. Every person who shall kill any moose, elk, mountain sheep or deer, for the purpose of obtaining the skin, hide, ham or hams of such animals, shall be guilty of a misdemeanor.

§ 2. **Deer.**—Every person who shall kill any spotted fawn shall be guilty of a misdemeanor. § 3. Every person who shall, between the 1st day of December in each year and the 1st day of August in the following year, or shall, between one hour after sunset and one-half hour before sunrise of any day, hunt, kill or destroy any deer, shall be guilty of a misdemeanor. Every person who shall kill any deer at any time, unless the carcass of such animal is used or preserved, or is sold for food, shall be guilty of a misdemeanor.

§ 4. **Hounds.**—Any person who shall hunt elk or deer with hounds, with intent to kill said elk or deer, shall be guilty of a misdemeanor.

§ 6. **Birds.**—Every person who shall, between the 1st day of December and the 1st day of September of the following year, kill or have in possession, except for breeding purposes, or sell or offer for sale, any grouse, pheasant, mongolian pheasant, quail or partridge, shall be guilty of a misdemeanor; provided, however, that it shall be unlawful to kill any prairie chicken, except during the months of July, August and September of each year. It shall be unlawful within the State of Oregon to kill any ring neck mongolian pheasant, or any of the various kinds of pheasants imported into this state by the Hon. O. N. Denny, or any quail, bob white or pheasant in that portion of the State of Oregon lying east of the Cascade Mountains.

§ 8. **Lawful Number.**—No person shall in one day kill or destroy a greater number than twenty of the hereinbefore enumerated birds.

§ 9. **Export.**—Every person or servant, agents, employes or operatives of any railroad, steamboat, express or other company or corporation, who shall transport or carry out of the state, or have in possession for the purpose of shipment, or carriage outside the State of Oregon, any of the birds named in the foregoing section, except for breeding or exhibition purposes, without written consent for the same having first been obtained, upon affidavit, from the Fish and Game Protector, shall be guilty of a misdemeanor.

§ 14. **Trout.**—Every person who shall, during the months of November, December, January, February and March of any year, catch any mountain lake, brook or speckled trout from any fresh water, shall be guilty of a misdemeanor. Every person who shall catch with any other device than hook and line any mountain or brook trout shall be guilty of a misdemeanor.

§ 15. **Wildfowl Methods.**—Every person who shall use any sink box on the Columbia River or any lake or river for the purpose of shooting wild ducks, geese, swan or other water fowl therefrom, at any time, shall be deemed guilty of a misdemeanor. § 16. Every person who shall use any batteries or swivel pivot gun, or any other gun than one, to be held in the hands and fired from the shoulder, either from the shore or on a boat, raft or or other device, on the Columbia River or on any lake or river, at any time, for the purpose of shooting wild ducks, geese, swan or other water fowl, shall be guilty of a misdemeanor. § 17. Every person who shall use or build any blind or other structure in any public lake or river, or in the Columbia River, more than 100 feet out from the shore or margin of such lake or river, for the purpose of shooting wild ducks, geese, swan or other water fowl therefrom, at any time, should be guilty of a misdemeanor. § 18. Every person who shall build or use any blind or other structure in any lake other than public lakes, more than 100 feet out from the shore or margin of such lake, for the purpose of shooting wild ducks, geese, swan or other water fowl therefrom, at any time, shall be guilty of a misdemeanor. § 19. Every person who shall build or use any blind or other structures in any lake, which lake is not wholly owned by himself, or his lessor or licensor, which stands more than 100 feet out from his own shore or margin of such lake, for the purpose of shooting wild ducks, geese, swan or other water fowls therefrom, shall be guilty of a misdemeanor. § 20. Every person who shall at any time between one hour after sunset and half an hour before sunrise fire off any gun, or build any fire or flash any light or powder or other inflammable substance upon the margin or in the vicinity of or upon any lake, pond, slough or other feeding grounds frequented by wild ducks, geese, swan or water fowl, with intent to thereby shoot, kill or disturb any of such water fowl, shall be guilty of a misdemeanor; provided, however, that it shall be lawful to shoot ducks and geese in or upon grain fields at any time to prevent the destruction of grain or growing crops; and provided further, that Curry county shall be exempt from the operation of this law.

§ 21. **Wildfowl Season.**—Every person who shall, between the 15th day of March and the 1st day of September of each year, kill or have in possession any wild swan, mallard ducks, wood ducks, widgeon, teal, spoonbill, gray, black, sprigtail or canvas-back duck, shall be guilty of a misdemeanor.

§ 25. **Robin, Meadow Lark, Other Birds.**—Every person who shall kill or destroy or have in his possession, except for breeding purposes, any nightingale, skylark, black thrush, gray singing thrush, linnet, goldfinch, greenfinch, chaffinch, bullfinch, red-breasted European robin, black starling, cross beak, Oregon robin or meadow lark, or mocking bird, shall be guilty of a misdemeanor. § 26. Every person who shall within the State of Oregon, at any time after the passage of this act, destroy or remove from the nest of any nightingale, skylark, black thrush, gray singing thrush, linnet, goldfinch, greenfinch, chaffinch, bullfinch, red-breasted European robin, black starling, cross beak or mocking bird, any egg or eggs of such bird, or have in possession, sell or offer for sale any such egg or eggs, or wilfully destroy the nests of such birds, shall be guilty of a misdemeanor. § 27. That every person who shall wilfully take, injure or destroy any sea gull shall be guilty of a misdemeanor.

§ 28. **Trespass.**—No person shall at any time enter into any standing or growing grain not his own, with intent to catch, recover, take or kill any bird or animal, nor permit any dog with which he shall then be hunting to do so for such purpose, without permission from the owner or the person in charge thereof. It shall be unlawful for any person to shoot upon or from the public highway.

SOUTH DAKOTA.

§ 1. Act of Feb. 4, 1893. **Buffalo, Elk, Deer, Antelope, Mountain Sheep.**—It shall be unlawful for any person or persons to kill, ensnare or trap in any form or manner or by any device whatever, or for any purpose, any buffalo, elk, deer, antelope, or mountain sheep from and after the passage and approval of this act until the 1st day of September, 1896.

§ 2366 (as amended by Act of Feb. 21, 1893). Political Code, Art. VII. **Prairie Chicken, Grouse, Snipe, Plover, Curlew, Wild Duck, Song Birds.**—It shall be unlawful for any person within this State to shoot or kill any prairie chicken, or pinnated grouse, or sharp-tailed grouse, or ruffed grouse, between the 1st day of January and the 1st day of September, or any wild duck, or snipe, or plover, or curlew, between the 15th day of May and the 1st day of September, or any song bird at any time. § 2367. It shall be unlawful for any person at any time or at any place within this State to shoot or kill for traffic any prairie chicken, wild duck, snipe, plover, or curlew, or for any person to shoot or kill during any one day more than twenty-five of said named birds, or for any one person, firm or corporation to have more than twenty-five of said named birds in his or their possession at any one time unless lawfully received for transportation.

Chap. 95, Laws 1890 (as amended by Act of Feb. 21, 1893). **Quail.**—It shall be unlawful for any person or persons to trap, ensnare or destroy by any means whatever any quail in this state for a period of five years from and after the 1st day of January, 1893.

§ 4. Act approved March 12, 1895. **Fish Seasons, Export.**—That it shall be unlawful to kill, take or have in possession any trout, bass, carp, shad or croppies taken or killed in any of the waters of this State during the months of October, November, December, January, February, March and April, or either of said months in any year. It shall be unlawful to sell or offer for sale at any time any trout or other food fish taken or killed in any of the waters of this State, or to ship or transport them out of the State.

TEXAS.

ART. 426. **Deer.**—It shall be unlawful to kill any wild deer between the 20th day of January and the 1st day of August.

ART. 426½. **Wild Turkey.**—It shall be unlawful to kill any wild turkey between the 15th day of May and the 1st day of September.

ART. 427. **Pinnated Grouse, Prairie Chicken.**—If any person shall kill any pinnated grouse (prairie chickens) in the months of March, April, May, June and July, he shall be deemed guilty of a misdemeanor.

ART. 428. **Quail, Partridge.**—If any person shall kill any quail or partridges in the months of April, May June, July, August and September, he shall be deemed guilty of a misdemeanor.

§ 1. Chap. 71, Laws 1891. **Birds of Plume.**—That if any person shall wilfully kill any seagull, tern, shearwater, egret, heron or pelican, or shall wilfully take from their nests or in any manner destroy any egg or eggs of any seagull, tern, shearwater, egret, heron or pelican, he shall be deemed guilty of a misdemeanor.

UTAH.

CHAPTER XCIV.

AN ACT FOR THE PROTECTION OF FISH, GAME AND BIRDS.

Be it enacted by the Legislature of the State of Utah:

§ 5. **County Wardens.**—At their first session after the passage of this act, the County Commissioners of each county of the state shall appoint a County Fish and Game Warden whose term of office shall be two years and until his successor shall be appointed and qualified.

§ 6. It shall be the duty of the County Warden to see that all laws of the state for the protection of fish and game are faithfully enforced in their respective counties, and for this purpose they are hereby given the same authority exercised by sheriffs and constables. It shall be the duty of the County Warden to report his official acts to the County Commissioners of his county annually.

§ 9. **Fishways and Guards.**—The owner or owners of any dam erected across any of the streams of this state, shall, if required by the County Fish and Game Warden, and under his direction, erect and maintain at all times at the expense of said owner or owners, suitable fishways to allow the free and uninterrupted passage of fish up and down the stream.

§ 10. It shall be unlawful for any person or persons operating any mill, factory, power plant or other manufacturing concern run by water power and having either head or tail races, without first furnishing and maintaining suitable screens or other device to prevent the fish from entering therein, said screens to be built and maintained under the direction of the County Warden and at the expense of said owner or owners, or operators of said mill, factory, power plant or other manufacturing concern; provided, that the woolen factory race at Provo, Utah County, this state, through which fish reach Spring Lake, be made an exception to these obligations.

§ 11. **Restrictions and Limits.**—It shall be unlawful for any person to take any fish, except carp, chubs, suckers and mullet from any of the waters of this state, by any means or device whatsoever, except by means of hook and line, commonly known as angling, and that only between the first day of July of each year and the fifteenth day of January following.

§ 12. It shall be unlawful for any person to sell, kill, destroy or have in his possession at any time, any fish except carp, chubs, sunfish or silver-side less than eight inches long, or any fish whatever that is taken unlawfully, or take, kill, destroy or have in his possession any trout or bass whatever at any time after the 14th day of January and before the 15th day of July following.

§ 13. It is hereby made unlawful for any person to take, kill or have in his possession any shad, catfish, whitefish, perch, rock-bass, crappie, rainbow trout, goldfish, silverfish or silver eels, for a period of three years after the passage of this act.

§ 14. It shall be unlawful for any person to kill or take any fish from the waters of the state by the use of any poison, deleterious or stupifying drug, giant powder, or quicklime or any explosive substance whatsoever. Any person found guilty of violating any of the provisions of this section shall be guilty of a felony.

§ 15. **Series Defined.**—It shall be unlawful for any person to take any fish except carp, chubs and suckers, from the public waters of the state, by the erection of any weir, dam, fence, wheel, basket, trap, net, seine, setline, sieve, spear or gun, or any other device whatsoever, which can and may be used for the unlawful catching of fish; provided, that for the purpose of catching carp, chubs, mullet or suckers, and these fish only, seines not more than two hundred yards long and twelve feet wide, with meshes not less than one and one-half inches square for fifty yards in the center, and meshes not less than two inches square in the wings thereof, may be used in Utah, Bear and Sevier lakes, and in Green and Grand rivers and in the lower and sluggish portions of the Bear, Weber and Jordan rivers between September 1st and May 15th following, both days inclusive; provided, that before any person shall use seines in the waters above mentioned such person shall secure the presence of either the county warden or his deputy, who shall be paid not to exceed the sum of $2 per day by the parties drawing the seines.

§ 16. It shall be unlawful to use seines within one-half mile from the mouth of any stream flowing into any lakes in the state, or the mouth of Spring Creek channels which lead into Utah lake, or within one-half mile from the mouth of any public stream or body of water connecting two other bodies of water, or anywhere within the confines of Spring Creek lake, which flows into Utah lake.

§ 17. Provides that all irrigation canals shall be protected by some device other than screens, so that no fish may enter same after May 15th of each year, and a failure to comply is a misdemeanor.

§ 18. Provides that the owner or operator of any reservoir shall provide and maintain at their expense, suitable screens to prevent fish from any waters of the state entering said reservoir.

§ 19. **Unlawful Appliances.**—All seines, nets, tackle, powder, explosives, lime, poisons, drugs and other means or devices for unlawfully taking or killing fish of any kind, found in the possession of any person who may be detected in unlawfully taking fish from any of the waters of the state, shall be seized by the officer making the arrest, and if used, or were about to be used, or intended to be used for the unlawful taking of fish, the same are hereby confiscated.

§ 20. **Deer and Elk.**—It shall be unlawful for any person to kill, wound, ensnare or trap within this State for the term of three (3) years from the passage of this act any elk, deer, buffalo, bison, antelope or mountain sheep, or any fawn or young of any of said animals; and the possession of the whole or any part of the carcass of any of said animals shall be prima facie proof of such unlawful taking or killing.

§ 21. It shall be unlawful for any person to pursue with any dog or dogs any of the animals mentioned in section 20 of this act.

§ 22. **Protection for Birds.**—It shall be unlawful for any person to kill, ensnare, net or entrap, or have in his possession within the state any partridge, pheasant, prairie chicken, or grouse after the first day of December and before the first day of August following of any year, or to rob or destroy the nest, eggs or young of any of the birds mentioned in this section.

§ 23. It shall be unlawful for any person to kill or have in his possession any sage-hen after the first day of March and before the first day of August following in any year.

§ 24. It shall be unlawful for any person to kill, ensnare, net, entrap at any time in any year any gull, owl, hawk, dove, lark, whip-poor-will, thrush, swallow, snowbird, bobolink, woodpecker or other insectivorous birds, except the English sparrow, or to rob or destroy the nest, eggs or young of any of said protected birds mentioned in this section.

§ 25. It shall be unlawful for any person to take, kill, wound, destroy, shoot at or have in his possession any wild goose, wild duck, snipe, brant or swan between the first day of April and the first day of October following, and it shall be unlawful for any person to use any gun larger than a ten (10) gauge while hunting for fowl or birds.

§ 26. **Quantities Stated.**—It shall be unlawful for any person to take, kill or have in his possession in any one day more than fifteen pounds of trout.

§ 28. It shall be unlawful for any person or persons at any time to ship or cause to be shipped, carried or transported out of the state any of the animals, birds or fish, or any part thereof mentioned in this act excepting carp, chubs, suckers and mullets.

§ 29. Any person who shall hereafter at any time within the state wilfully kill, wound, ensnare, trap, shoot at or have in his possession any Mongolian or Chinese pheasant, English pheasant, pinnated grouse or quail except the quail found in Southern Utah, commonly called the California quail, shall be guilty of a misdemeanor.

§ 30. **Penalties Provided.**—Any person violating any of the provisions of this act, other than the provision of section 14, shall be guilty of a misdemeanor, and shall be fined not less than $5 nor more than $300, or imprisonment not less than five days nor more than thirty days, or both, at the discretion of the court. All fines and forfeitures collected under the provisions of this act shall be paid into the county treasury of the respective counties.

§ 31. Chapter LXXVIII of the Session Laws of 1894 entitled "An act to provide for the protection of fish and game and for the appointment of Territorial and county commissioners," and all other acts and parts of acts in conflict herewith are hereby repealed.

WASHINGTON.

§ 1. Act. of Feb. 6, 1890. **Use of Dogs—Hides or Horns.**—That it shall be unlawful to hunt or chase deer with dogs. § 2. That it shall be unlawful to hunt deer, mule deer, caribou, elk, mountain sheep or goats for the purpose of securing their hides or horns.

§ 1. Chap. 54, Laws 1888. **Elk, Moose, Deer, Mountain Sheep, Mountain Goat.**—That every person who shall between the 1st day of January and the 15th day of August kill any elk, moose, deer, fawn, mountain sheep or mountain goat, shall be deemed guilty of a misdemeanor. Every person who shall kill any elk, moose, deer, fawn, mountain sheep or goat, at any time, unless the carcass of such animal is used or preserved for food; or shall, between the 1st day of January and the 15th day of August, sell any hides or horns of same, shall be deemed guilty, etc.

§ 2. **Hunting Elk and Moose With Dogs.**—Every person who shall hunt any elk or moose with dog or dogs, at any time, except during the months of October, November and December, shall be guilty of a misdemeanor.

§ 3. **Wildfowl** —Every person who shall, between the 1st day of April and the 15th day of August, kill any wild swan, mallard duck, wood duck, widgeon, teal, butter-ball, spoon-bill, blue-bill, red-head, gray duck, black duck, sprigtail or canvas-back duck, shall be guilty of a misdemeanor. § 11. Every person who shall kill any mallard duck, widgeon, teal, butter-ball, spoon-bill, wood duck, gray duck, black duck, blue-bill, red-head, sprigtail or canvas-back duck at any season of the year, between the hours of 8 o'clock p. m. and 5 o'clock a. m., shall be guilty of a misdemeanor. § 13. Every person who shall use any sink-box, floating blind, rafts, sneak boat, punt or any other device for approaching any of the water fowl mentioned in this chapter, while the same are resting on the waters of this State, shall be guilty of a misdemeanor; provided, that nothing in this chapter shall be construed to prevent the shooting of any of the water fowl mentioned therein from shore blinds or over decoys with any gun which is fired from the shoulder of the shooter.

§ 4. **Grouse or Pheasant, Sage Hen.**—Every person who shall, between the 1st day of January and the 1st day of August of each year, kill any mountain grouse, blue or dusky grouse, ruffed grouse or pheasant, pintail grouse, or prairie chicken or sage hen, shall be guilty of a misdemeanor.

§ 1. Act of March 9, 1891. **Killing Feathered Game for Sale.**—It shall be unlawful to kill any feathered game for the market or sale in any month except the month of December. § 2. Such game shall be of the several kinds, as follows: Swan, geese, brants, sand-hill cranes, grouse, pheasants, partridges, prairie chicken, snipe, and all the various and different kinds of ducks.

§ 5. **Quail, Mongolian Pheasants.**—That it shall be unlawful to kill quail and golden, silver, China of Mongolian pheasants for the period of five years after this act becomes a law. [March 9, 1891.]

§ 1. Act of March 6, 1891 (as amended 1895.) **Trout.**—Every person who shall during the months of November, December, January, February and March of each year take any trout, mountain trout, shall be guilty of a misdemeanor. Every person who shall take any of the food fishes implanted in the creeks, rivers, lakes or bays, except for propagating the same, for a period of three years after the same shall have been implanted, shall be guilty of a misdemeanor.

§ 1. Act of Feb. 11, 1890. **Salmon.**—It shall not be lawful to take salmon in the Columbia River or its tributaries by any means whatever between the 1st day of March and the 10th day of April, or between the 10th day of August and the 10th day of September, or in any rivers and bays of the State, or the Columbia River, between the hours of 6 o'clock p. m. on each and every Saturday and 6 o'clock in the afternoon of the following Sunday.

WYOMING.

CHAP. 93, LAWS 1895.

§ 6. **Game and Fish may be Taken only for Immediate Use as Food.** —The wanton destruction or the wasting of the game and fish of this State during any period of time when the taking or capture of such game or fish is permitted is hereby prohibited and declared a misdemeanor; and any person who shall at any time take, capture or destroy any game or fish in excess of the number or quantity thereof which he can immediately use for food purposes shall be deemed guilty of a misdemeanor.

§ 7. **Trout, Land-locked Salmon, Grayling.**—No person shall catch or take, or have in his possession, from any lake, river or stream of this State by any means whatever, any speckled trout, land-locked salmon, grayling or California trout, during the months of November, December, January, February, March, April and May, or either of said months in any year.

§ 8. **Size of Trout and Bass.**—It shall be unlawful to kill or destroy, or have in possession for any purpose at any time, any trout or black bass less than six inches in length, taken from any of the waters of this State.

§ 12. **Birds.**—No person shall kill, net or trap within this State any snipe, green shank, tatler, godwit curlew, avoset, or other wader or plover, nor quail, lark, whip-poor-will, finch, thrush, snow-bird, turkey buzzard, robin or other insectivorous birds, except that partridges, pheasants, prairie chickens, prairie hens or grouse may be shot from August 15th to November 1st of each year, and sage chickens may be shot from August 1st to October 15th of each year.

§ 13. **Wildfowl.**—No person or persons shall ensnare, net or trap within this State any wild duck or wild goose at any time. There shall be established from the 1st day of September until the 1st day of May an open season, in which ducks, brant, geese and swans may be shot, killed or taken by means of gun shot; and it shall be illegal to kill any of the above-named waterfowl by any other means, or during any other period than above specified.

§ 14. **Big Game.**—It shall be unlawful to pursue, hunt or kill any deer, elk, moose, mountain sheep, mountain goat or antelope at any time except during the months of September, October and November in each year, during which months the males of such animals may be killed or hunted under the conditions and restrictions imposed by this section. It shall be unlawful at any time whatever to kill or capture any of the above-named animals mentioned in this section by means of any pit, pitfall or trap. Any person may, during the period permitted and prescribed in this section, pursue, hunt and kill any of the males of the animals mentioned in this section for the purpose only of supplying himself with food, but not for speculative purposes or wantonly. It is hereby declared to be unlawful for any Non-resident of this State to hunt, kill or pursue any of the male animals permitted by this section to be hunted, killed or pursued herein, without first having procured a license therefor so to do from a justice of the peace of the county wherein said animals are to be hunted. The justices of the peace of this State are hereby authorized and directed to issue such licenses upon the payment of $20 for each license, which shall be good in their county. Such license shall permit such non-resident to pursue, hunt or kill any of the males of the animals mentioned in this section during the months of September, October and November of the current year for the purpose of supplying himself or his family with food during such period.

§ 15. **Capturing for Export.**—It shall be unlawful at any time to capture or pursue, for the purpose of capturing any of the animals mentioned in section 14 of this act, of whatever age, for the purpose of selling or disposing of the same, or of shipping the same out of the State, except by express permission in writing of the State game and fish warden, and then only for the purpose of supplying public parks, zoological gardens or places of public amusement.

§ 16. **Bison, Buffalo, Beaver.**—No person shall kill, wound, ensnare or trap any bison or buffalo within the State of Wyoming for a period of ten years from and after the approval of this act. No person shall kill or wound, ensnare or trap, any beaver or kitten beaver within the State of Wyoming for a period of five years from and after the approval of this act.

§ 17. **Hides, Horns.**—It shall be unlawful for any person or persons to purchase, or obtain by barter, any green, tanned or untanned hide or hides or horns of any of the animals mentioned in § 14.

§ 18. **Transportation.**—It shall be unlawful for any railway, express company, stage line or other public carrier, or any of their agents or employes, or other person or persons, to receive or have in their possession for transportation any carcass or part of carcass, hides, tanned or untanned, or horns, of any of the animals mentioned in § 14 of this act, or to transport the same after the passage of this act; except that nothing in this act shall prevent shipping or transporting in any manner mounted heads or stuffed birds or animals to any point in or out of the state.

§ 20. **Dogs.**—It shall be unlawful for any person or persons to use dogs for the purpose of running or coursing deer, antelope, elk or mountain sheep.

YELLOWSTONE NATIONAL PARK.
ACT OF CONGRESS, MAY 7, 1894.

§ 4. That all hunting or the killing, wounding or capturing at any time of any bird or wild animal, except dangerous animals, when it is necessary to prevent them from destroying human life or inflicting an injury, is prohibited within the limits of said park; nor shall any fish be taken out of the waters of the park by means of seines, nets, traps, or by the use of drugs or any explosive substances or compounds, or in any other way than by hook and line, and then only at such seasons and in such times and manner as may be directed by the Secretary of the Interior.

YELLOWSTONE NATIONAL PARK RULES.

1. It is forbidden to remove or injure the sediments or incrustations around the geysers, hot springs or steam vents; or to deface the same by written inscription or otherwise, or to throw any substance into the springs or geyser vents; or to injure or disturb, in any manner, any of the mineral deposits, natural curiosities or wonders within the Park. 2. It is forbidden to ride or drive upon any of the geyser or hot spring formations, or to turn loose stock to graze in their vicinity. 3. It is forbidden to cut or injure any growing timber. Camping parties will be allowed to use dead or fallen timber for fuel. 4. Fires shall be lighted only when necessary, and completely extinguished when not longer required. The utmost care should be exercised at all times to avoid setting fire to the timber and grass. 5. Hunting, capturing, injuring or killing any bird or animal within the Park is prohibited. The outfits of persons found hunting or in possession of game killed in the Park will be subject to seizure and confiscation.

THE BRITISH PROVINCES.

BRITISH COLUMBIA.

GAME PROTECTION ACT, 1895.

§ 3. **Cow Elk and Moose, Ewe of Sheep, Hen Pheasant, Quail, English Partridge and Other Birds.**—None of the following animals or birds shall be killed at any time, viz.: Cow wapiti (commonly known as elk), cow moose, ewe or lamb of the mountain sheep, fawn, English blackbird, chaffinch, hen pheasant, linnet, skylark, thrush, robin, all species of quail, English partridge, and gull, or any bird known by any of these names, except as regards robin, as is provided in sub-section (*c*) of § 15. § 4. It shall be unlawful to take or destroy in any manner any bird which subsists principally on noxious insects.

§ 5. **Wildfowl in Victoria Harbor.**—It shall not be lawful at any time of the year to shoot any wildfowl or discharge a firearm within that part of Victoria Harbor to the north of a line drawn from Shoal Point, in the City of Victoria, to Work Point, in the District of Esquimalt.

§ 7. **Export.**—No person shall at any time purchase or have in possession, with intent to export, or cause to be exported or carried out of the limits of this province, any or any portion of the animals or birds mentioned in this act, in their raw state; and this provision shall apply to railway, steamship and express companies. Provided that it shall be lawful for any person having a license under § 20 of this act to export, or cause to be exported or carried out of the province, the heads. horns and skins of such animals mentioned in § 21 of this act as having been legally killed by such license-holder; provided that the provisions of this section shall not apply to bear or beaver.

§ 8. **Cock Pheasants.**—It shall not be lawful on the mainland of British Columbia to shoot, capture, trap or by any means destroy cock pheasants.

§ 9. **Imported Game.**—No person other than the importer, on his own property held as a private preserve, shall kill any game birds or animals hereafter imported for acclimatization purposes and distributed in any part of the province, until such time as the lieutenant-governor shall appoint.

§ 12. **Forbidden Fishing Methods.**—No person shall use any explosive or poison, net, seine, drag net or other device other than hook and line, nor use salmon roe as bait for the purpose of taking or capturing trout.

§ 13. **Deer Hides.**—It shall be unlawful to kill deer at any time for their hides alone in any portion of this province.

§ 14. **Dogs for Deer.**—No person shall at any time hunt deer with dogs.

§ 15. **Close Seasons.**—Except as hereinafter provided, it shall be unlawful to kill between sunset and sunrise throughout each year, or within the periods and localities hereinafter limited:

(*a.*) To the east of the Cascade Range—Any variety of grouse, including willow, ruffed, blue, prairie hen, prairie chicken, ptarmigan, Franklin's or fool hen and meadow lark, from the 16th day of November to the 31st day of August, inclusive. Wild duck of all kinds, bittern, plover and heron, from the 1st day of January to the 31st day of August, inclusive.

(*b.*) Throughout the Province—Caribou, deer, wapiti (commonly known as elk), moose, hare, mountain goat and mountain sheep, from the 1st day of January to the 31st day of August, inclusive.

(*c.*) West of the Cascades—Any variety of grouse, including willow, ruffed, blue, prairie hen or prairie chicken, ptarmigan, Franklin's or fool hen and meadow lark, from the 1st day of December to the 30th day of September, inclusive. Wild ducks of all kinds, plover, bittern and heron, from the 1st day of March to the 31st day of August, inclusive. Provided that the birds known in this province as "robins" may be destroyed in an orchard or garden at any time between the 1st day of June and the 1st day of September.

(*d.*) On Vancouver Island—Cock pheasants, from the 1st day of February to the 30th day of September, inclusive.

§ 21. **Non-Residents.**—No person who is not domiciled in this province, other than officers and men in Her Majesty's regular army and navy, or in the permanent corps of Canadian militia for the time being in actual service in this province, shall at any time hunt, kill or take any of the animals mentioned in the next succeeding section, without being authorized thereto by license. § 20. Such license may, on payment of a fee of $50, be granted by any government agent in the province to any person who shall apply to him therefor, and shall be valid only for that shooting season for which the same has been issued; and such license shall in no case give a right to the holder thereof to kill, in addition to the birds mentioned in this act which may be killed, more than ten deer, five caribou, five mountain sheep, five mountain goats, two bull wapiti or elk, and two bull moose.

§ 21. **Lawful Number.**—No one shall, during any one year or season, kill more than ten deer, five caribou, five mountain sheep, five mountain goats, two bull wapiti or elk, or two bull moose.

§ 23. **Wildfowl.**—No person shall catch, kill or take more than 250 ducks during any one season; nor shall anyone use, for taking or killing wild ducks of any kind, or geese, contrivances described or known as batteries, swivel guns or sunken punts in non-tidal waters.

§ 25. **Trespass.**—No person shall at any time enter into any growing or standing grain not his own with sporting implements about his person, nor permit his dog or dogs to enter into such growing or standing grain without permission of the owner or occupant thereof; and no person shall at any time hunt or shoot upon any enclosed land of another after being notified not to hunt or shoot thereon.

CANADIAN DUTY ON GUNS AND RODS.

The duty is 30 per cent of the appraised value.

MEMORANDUM. NO. 492 B.

CUSTOMS DEPARTMENT, OTTAWA, July 4, 1891.—*Collector of Customs:* I am now instructed by the Honorable the Minister of Customs to authorize you to accept entry and duty on the guns, fishing rods and other equipments of parties visiting Canada for sporting purposes, with the condition that the duty so paid will be refunded on proof of exportation of the same within a period of two months from the date of entry.

J. JOHNSON, Commissioner of Customs.

MEMORANDUM. NO. 556 B.

CUSTOMS DEPARTMENT, OTTAWA, June 14, 1892.—*Collector of Customs:* Referring to Memo. No. 492 B, of July 4, 1891, *re* sportsmen's guns, etc. * * * You will also inform all entering their outfits that if they expect under the terms of the memorandum a refund of the duty paid, it can only be granted on condition that the claim bears a Canadian custom officer's certificate of identification and the usual evidence of exportation.

W. G. PARMELEE, Commissioner of Customs.

NORTHWEST TERRITORIES.

§ 2. The Game Ordinance of 1893. **Game Seasons.**—No elk, moose, caribou, antelope, deer or their fawn, mountain sheep or goat, shall be killed between the 1st day of February and the 1st day of September. (1.) No person shall take more than six head of the aforesaid animals in one season except for food for himself or family. § 3. No person shall kill: (1.) Any buffalo; (2.) any grouse, partridge, pheasant, or prairie chicken between the 1st day of January and the 1st day of September; (3.) any kind of wild duck between the 15th day of May and the 23d day of August; (4.) any plover, snipe and sandpiper between the 1st day of January and the 1st day of August.

§ 5. **Wildfowl Methods.**—None of the contrivances for the taking or killing of the wildfowl, known as swans, geese or ducks, which are described as swivel guns, batteries, sunken punts, or night lights, shall be used at any time.

§ 11. **Export.**—No person shall export out of the limits of the Northwest Territories any grouse, partridge, pheasant, prairie chicken, elk, moose, caribou, antelope, or their fawn.

Non-Residents.—No person, who is not a resident of the Territories, shall kill any of the aforesaid animals or birds unless he has obtained from the Lieutenant-Governor-in-Council a license on payment of $5, which license shall expire on the 15th day of May in each year. (a.) Provided that a license may be granted to a guest of any resident free for a term not exceeding five days.

§ 1. General Regulations, July, 1889. **Fish.**—No one shall catch any pickerel (dore) between the 15th day of April and the 15th day of May, both days inclusive. § 4. No one shall catch any speckled trout (*Salvelinus fontinalis*) between the 1st day of October and the 1st day of January.

SHOTGUN CHARGES FOR GAME.

The following loads are those adopted by the Union Metallic Cartridge Co. in their "New Club" loaded shells for game. Black powder is used.

10-GAUGE.					12-GAUGE.			
LOAD NO.	ADAPTED TO SHOOTING.	POWDR DRAMS	SHOT, OUNCE.	NO.	LOAD NO.	POWDR DRAMS	SHOT, OUNCE.	NO.
240	Woodcock........................	4	1⅛	10	**110**	3¼	1	10
249	Woodcock and Snipe.	4	1⅛	9	**119**	3¼	1	9
269	Snipe..................................	4¼	1⅛	9	**129**	3¼	1⅛	9
248	Quail..............	4	1⅛	8	**118**	3¼	1	8
268	Quail and Prairie Chicken ...	4¼	1⅛	8	**128**	3¼	1⅛	8
278	Inanimate Targets......	4¼	1¼	8	**78**	3¼	1¼	8
288	Prairie Chicken..................	4½	1⅛	8	**138**	3½	1⅛	8
298	Live Pigeons, etc................	4½	1¼	8	**148**	3½	1¼	8
267	Ruffed Grouse............	4¼	1⅛	7	**127**	3¼	1⅛	7
287	Ruffed Grouse and Teal........	4½	1⅛	7	**137**	3½	1⅛	7
266	Prairie Chicken and Ruffed Grouse	4¼	1⅛	6	**126**	3¼	1⅛	6
296	Bluebill and Pintail.......... ...	4½	1¼	6	**136**	3½	1⅛	6
265	Mallard.	4¼	1⅛	5	**125**	3¼	1⅛	5
285	Mallard.	4½	1⅛	5	**135**	3½	1⅛	5
264	Mallard.	4¼	1⅛	4	**124**	3¼	1⅛	4
284	Mallard and Red-head...........	4½	1⅛	4	**134**	3½	1⅛	4
294	Redhead	4½	1¼	4	**154**	3¾	1⅛	4
293	Canvas-back......................	4½	1¼	3	**153**	3¾	1⅛	3
283	Canvas-back......................	4½	1⅛	3				
322	Turkey..............................	5	1⅛	2	**152**	3¾	1⅛	2
352	Goose and Brant..................	5	1¼	BB				

UNION PACIFIC,

"THE OVERLAND ROUTE."

Information regarding the territory traversed by the UNION PACIFIC SYSTEM, or the various resorts along the line, will be cheerfully furnished on application to any representative of the UNION PACIFIC at the agencies named below. They will also make the necessary arrangements for transportation when so desired:

S. A. HUTCHISON, General Traveling Passenger Agent, Omaha, Neb.

ALBANY, N. Y.—23 Maiden Lane.
BOSTON, MASS.—292 Washington Street.
BUFFALO, N. Y.—210 Ellicott Square.
BUTTE, MONT.—50 North Main Street.
CHEYENNE, WYO.
CHICAGO, ILL.—191 So. Clark Street.
CINCINNATI, OHIO.—Room 35 Carew Building.
COUNCIL BLUFFS, IOWA.—U. P. Transfer.
DENVER, COLO.—No. 941 Seventeenth Street.
DES MOINES, IOWA—401 Walnut Street.
DETROIT, MICH.—67 Woodward Avenue.
HELENA, MONT.—56 North Main Street.
KANSAS CITY, MO.—1000 Main Street.
LONDON, ENG., 122 Pall Mall.
LOS ANGELES, CAL.—223 So. Spring Street.
NEW YORK CITY—287 Broadway.
OAKLAND, CAL.—1010 Broadway.
OGDEN, UTAH—Union Depot.
OMAHA, NEB., 1302 Farnam Street.
PHILADELPHIA, PA.—Room 3, No. 20 So. Broad Street.
PITTSBURG, PA.—1016 Carnegie Building.
PORTLAND, ORE.—135 Third Street.
ST. JOSEPH, MO.—Chamber of Commerce.
ST. LOUIS, MO.—213 North 4th Street.
SALT LAKE CITY, UTAH—201 Main Street.
SAN FRANCISCO, CAL.—No. 1 Montgomery Street.
SEATTLE, WASH.—618 First Avenue.
SIOUX CITY, IOWA—506 Fourth Street.
TACOMA, WASH.—907 Pacific Avenue.
YOKOHAMA, JAPAN—No. 4 Water Street.

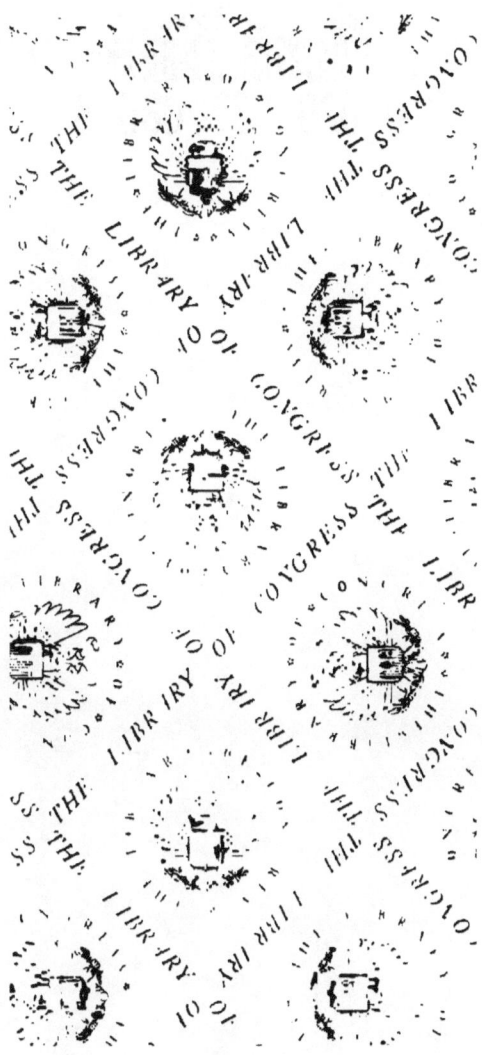

www.ingramcontent.com/pod-product-compliance
Lightning Source LLC
Chambersburg PA
CBHW022030190326
41519CB00010B/1645